TV TUBE SYMPTOMS and TROUBLES

by
ROBERT G. MIDDLETON

HOWARD W. SAMS & CO., INC.
THE BOBBS-MERRILL CO., INC.
INDIANAPOLIS · KANSAS CITY · NEW YORK

SECOND EDITION

SIXTH PRINTING—1971

Copyright © 1961 and 1966 by Howard W. Sams & Co., Inc., Indianapolis, Indiana 46206. Printed in the United States of America.

All rights reserved. Reproduction or use, without express permission, of editorial or pictorial content, in any manner, is prohibited. No patent liability is assumed with respect to the use of the information contained herein.

International Standard Book Number: 0-672-20476-2
Library of Congress Catalog Card Number: 66-24118

PREFACE

It is a well-known fact that most TV receiver troubles are caused by defective tubes. In trying to relate trouble symptoms to specific tubes, however, there is often very little correlation between tube types used in different makes of receivers, or even between different models of the same make. Therefore, to decide which tube or tubes could be the cause of a specific trouble, it becomes necessary to relate the symptom to the tube or tubes which perform a specific function.

Thus, this book serves a dual purpose. The first section explains the functions of each section in a TV receiver, and describes typical troubles which can be attributed to defective tubes in these sections. Section 2 pictures and describes over 75 trouble symptoms, providing a handy reference guide which can be used to quickly pinpoint defective TV receiving tubes.

Continued interest in TV tube symptoms and troubles has necessitated this second edition of this book. You have asked questions concerning tube testing; this topic is now given appropriate attention in Section 1. Color TV has become popular, and pointers have been included concerning tube troubles in the black-and-white section of color receivers. There is also a trend to hybrid receivers—that is, receivers containing part transistors and part tubes. Guiding notes have been provided on this topic. UHF has grown in popularity, and some mention of tube troubles in UHF tuners has been included.

While it would be impractical to include every cause and remedy for every symptom, you'll find that this book will help you correct virtually any TV trouble caused by a defective tube. If tube replacement does not cure the defect, further troubleshooting with test equipment and a schematic will be required. But if this book shows you how to cure up to 80% of all TV troubles more quickly—and hence, more profitably—its purpose will have been accomplished.

ROBERT G. MIDDLETON

TABLE OF CONTENTS

Section 1

General Introduction 7

    Functional Block Diagram—Tube Complement and Type Variations—Symptoms Related to Tube Functions—Variations in Trouble Symptoms—Hybrid Receivers—Color Receivers—UHF Tuner—Tube Testing—Other Helpful Hints—General Chart of Symptoms

Section 2

Picture-Tube Displays 49

    Picture Troubles 51
    Sync Troubles 65
    Size and Linearity Troubles 78
    Raster Troubles 88
    Sound Troubles 94

SECTION 1

GENERAL INTRODUCTION

The greatest majority of TV receiver troubles are caused by defective tubes. Hence, if we learn to recognize the picture and sound symptoms for the various defects, we can pinpoint the specific tube at fault. Although some receivers use more tubes than others, and there is considerable variation in circuit configurations, the basic principles are the same for all sets. That is, certain basic functions must be performed, and once you are familiar with the fundamental purpose of each section, you should have no trouble in diagnosing tube defects in any TV receiver.

Throughout this book, all symptoms are discussed in relation to the tubes which can cause them. Naturally, if tube substitution does not remedy the situation, defects in other circuit components are implied. Hence, this book will not only serve as a guide for tube defects, but will also help in locating the circuit which needs further troubleshooting.

FUNCTIONAL BLOCK DIAGRAM

The block diagram of a typical large receiver is given in Fig. 1. Other receivers may have more or fewer stages, de-

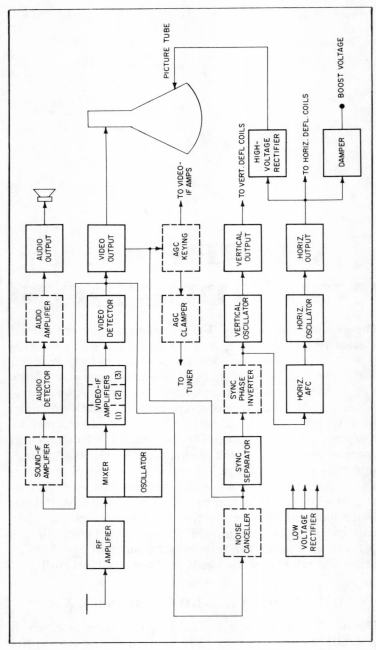

Fig. 1. Block diagram of a typical TV receiver.

pending on their intended application and price range. The blocks shown in dotted lines are the stages most commonly omitted from less expensive sets.

Here is how a typical TV receiver works: The signal from the station arrives at the antenna, along with all other station signals in the area. The RF amplifier selects only the desired signal (rejecting all others), amplifies it, and couples it to the mixer. In the mixer the station signal and a signal from the local oscillator are beat together (heterodyned) to produce a lower, or intermediate, frequency. The intermediate-frequency (IF) amplifiers are tuned to pass the IF frequencies—that is, a specific band approximately four megacycles wide—and to reject all others. This wideband response is obtained by using various means of coupling between the IF-amplifier stages. Although these stages are commonly termed the video-IF amplifiers, actually the composite signal which contains sound, sync, and video is present at this point in all modern receivers. Three stages of video IF are shown in Fig. 1; however, some receivers have four and others, two; in one instance, only one video-IF amplifier is employed.

The video detector detects, or demodulates, the video-IF signal, so that at its output there is a DC voltage that varies in step with the original televised signal. This signal also includes the sync pulses and the sound information. Since the video carrier and the sound carrier are separated by 4.5 mc in the IF signal, the two beat together in the detector to produce a 4.5-mc signal which contains all the sound information.

The video signal at the detector output is further amplified by the video-output tube and then applied to the picture tube, where it varies the intensity of the beam to produce light and dark spots corresponding to the original telecast scene. Only a single video-output tube is shown in Fig. 1; however, in some receivers two stages are employed.

The sound signal at the detector output is coupled to a sound-IF amplifier tuned to 4.5 mc. Here the signal is amplified and then applied to the audio detector. The audio detector produces a voltage, at the audio frequency, which varies in step with the original sound signal. The audio-amplifier and audio-output tubes amplify this audio signal to the level required to drive the speaker. In some receivers the sound-IF amplifier or the audio amplifier, or both, may be omitted.

A portion of the signal from the video-output tube is coupled to the AGC keying tube, which samples its amplitude and produces a bias voltage for the video-IF and RF amplifiers. This bias voltage controls the gain of these stages so that the output signal is fairly constant, even though the input signal varies in strength. In the keyed AGC system, the amplitude of the video signal is checked during the horizontal sync-pulse time only. The reason is that the horizontal-sync pulse is always transmitted at the same level, whereas the video-signal level will vary for light and dark scenes. Thus, keying the AGC voltage to the horizontal-sync pulse provides a constant check point.

Other systems may not be keyed; instead, the entire signal is used as a reference. This stage is then called the AGC amplifier. In still other receivers, no tubes are employed for AGC—it is provided by a combination of resistors and capacitors.

An AGC clamper (omitted in many receivers) is also shown in Fig. 1. This stage removes the AGC voltage from the RF amplifier and allows the amplifier to operate at full gain until the signal reaches a predetermined level, then normal AGC voltage is applied to the amplifier.

Now we have the video (picture) signal displayed on the screen of the picture tube and the sound emitting from the speaker, and we have also provided a means of controlling the gain of the video-IF and RF amplifiers so the signal will have a fairly constant amplitude. However, the signal will remain in one spot. To display a picture on the screen, this spot must be moved (scanned) across the screen from left to right, then quickly returned to the left side for the next line. The period when the beam moves from left to right is called the *trace*, and from right to left, the *retrace*. (The retrace portion is not visible on the screen.) Each line is directly below the previous one. When the bottom of the screen is reached, the beam is returned to the top of the picture to start over again. This time each horizontal trace line is located between the two lines from the previous time. This principle is termed *interlaced* scanning. That is, each time the picture is scanned, the horizontal trace lines are interlaced with the scan lines of the previous field.

The horizontal oscillator generates the signal that moves the beam from left to right on the screen. Likewise, the vertical oscillator generates the signal which moves the beam from the top of the screen to the bottom. The vertical- and horizontal-output tubes amplify the two signals and apply them to the deflection coils (around the neck of the picture tube). This action is independent of the received signal; therefore, even with no incoming signal—for example, on an unused channel—the screen will be lit. The lighted area caused by the scan lines is called the *raster*, and it should not be confused with the picture. The picture is the video information displayed on the screen to produce the varying shades of black and white. The raster is the scan lines with or without a picture. Thus, it is possible to have a picture even though there is no raster! In other words, the video information is being presented to the picture tube—the video stages are functioning—but no picture is seen because the screen is blacked out.

The video signal modulates (varies) the intensity of the beam as it is swept back and forth across the screen. This creates the light and dark portions of the picture. But before a viewable picture resembling the original scene can be obtained, the light and dark areas on the screen must correspond exactly with the same light and dark areas in the original scene. Thus, the vertical and horizontal oscillators in the receiver must not only operate at precisely the same frequency at which the original scene was scanned at the transmitter; they must also be in step (synchronized) so that each vertical field and each horizontal line starts at the same time as the original. This is the purpose of the sync pulse in the transmitted signal.

The video signal at the output of the video-output tube in Fig. 1 is coupled to the sync separator. The sync separator separates the horizontal- and vertical-sync pulses from the video signal and couples them to their respective oscillators, where they serve as a trigger to start each field or each line at exactly the right instant. A noise canceller is often added to prevent any noise pulses in the signal from feeding through and prematurely triggering the oscillators. The sync-phase inverter, when used, reverses the polarity of the signal so it will be of the proper phase to trigger the oscillators.

Notice that the sync pulses do not trigger the horizontal oscillator directly, but are fed to the horizontal AFC (automatic frequency control) tube first. This tube provides a correction voltage which keeps the oscillator frequency in step with the transmitted signal modulations. The basic purpose of the AFC tube is to improve the immunity of the circuit to noise, which might be present if the sync pulses were fed directly to the horizontal oscillator.

During horizontal retrace, when the beam quickly returns from the right to the left side of the raster, undesired oscillations are set up in the horizontal-output transformer. The purpose of the damper tube is to squelch, or damp, these oscillations. It also permits the deflection-coil current to decay at a uniform rate, so that the left side of the raster will be linear—that is, not distorted. The damper action results in a boost voltage which is higher than the normal B+ voltage. After being filtered, the boost voltage is used as the B+ for some of the stages.

The pulse from the horizontal-output tube also serves another purpose. It is stepped up by the horizontal-output transformer and then applied to the high-voltage rectifier tube, where it is rectified, or changed to DC, and applied to the picture tube. Here, the high voltage accelerates the beam so that it strikes the fluorescent face of the tube with sufficient force to cause the screen to glow.

The low-voltage rectifier changes the AC line voltage to the DC B+ voltages required for operation of the tubes. Often the line voltage is stepped up by a power transformer before it is rectified.

TUBE COMPLEMENT AND TYPE VARIATIONS

Some receivers have as few as 10 tubes; others have 30 or more. Generally the fewer the tubes, the more semiconductor diodes (Fig. 2), multisection tubes, and multipurpose tubes utilized. Thus, a typical 10-tube receiver contains several dual-section IF tubes, and one with 11 tubes

Fig. 2. Semiconductors used in TV receivers.

employs six semiconductor diodes. On the other hand, a receiver with 21 tubes may use single-section IF tubes and only one semiconductor diode (the UHF mixer).

Nor will the same tube type serve the same function in different makes of receivers. For example, one receiver may use a 6BA8 as a second-IF and horizontal oscillator, while another uses it as a video amplifier and phase splitter. Although the RF amplifier is a 2BN4 in one receiver, we find a 4BS8 performing this function in another receiver. So when in doubt, look for the tube-layout diagram on the cabinet or chassis, or in the receiver service notes. (See Fig. 3.)

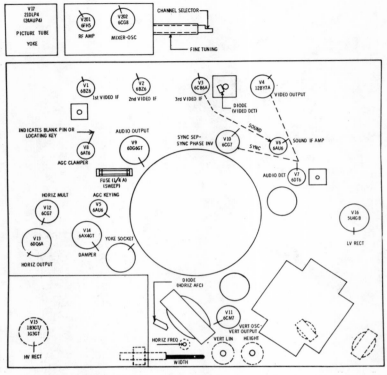

Fig. 3. A typical tube-placement chart.

These diagrams show where each tube belongs, and its function and type. They will help you find the location of a specific stage, and are valuable should you remove more than one tube at a time and forget which one goes where!

Tubes with the same basing (for example, 9-pin miniatures) can accidentally be plugged into the wrong socket. As a result, the tube or other circuit components could be damaged. On the other hand, it is often permissible to substitute certain tube types for other types, provided the two have similar characteristics and the same internal connections.

Separate sync-separator and noise-canceller functions were illustrated in Fig. 1, but they are often combined in a single pentagrid tube such as the 6CS6. In another popular version, both functions are accomplished in a 6BU8 dual-pentode tube which also provides AGC keying.

SYMPTOMS RELATED TO TUBE FUNCTIONS

In general, failure of a tube performing a certain function will cause a definite picture or sound symptom. (This assumption overlooks the possibility of interaction, to be discussed later.) In the following we will discuss and show the most common troubles associated with the various sections of the receiver. The troubleshooting chart at the end of this section gives a more comprehensive listing of the troubles associated with each tube. The normal picture is given in Fig. 4 for comparison.

Fig. 4. A normal picture.

Video Troubles

Failure of the RF amplifier often results in no picture, and noise (snow) is usually visible in the raster. (Strong signals are able to feed through a dead RF-amplifier stage and produce a picture, but it will be weak or snowy.) The effect on the sound varies from apparently no effect to complete loss, depending on the signal strength.

If the local oscillator fails, there will be no picture or sound. However, snow will appear in the raster, as seen in Fig. 5. Likewise, a dead mixer stage will result in no picture or sound, but snow will be present in some instances.

Failure of a video-IF amplifier tube will usually produce loss of picture and sound (Fig. 6), or weak picture and

Fig. 5. Snowy raster.

Fig. 6. Loss of picture, no snow.

Fig. 7. A weak picture.

sound. However, a snowy raster will sometimes be produced when the first video-IF amplifier fails. Likewise, low emission in a video-IF amplifier tube will produce a weak picture, as shown in Fig. 7. A gassy video-IF amplifier tube will cause a negative picture—that is, the white areas are reproduced as black, and the black areas as white. A negative picture is usually accompanied by poor or no vertical sync (Fig. 8) or poor horizontal sync (Fig. 9).

Fig. 8. A *negative picture with loss of vertical sync.*

Fig. 9. A *negative picture with loss of horizontal sync.*

Fig. 10. *Excessive contrast.*

In general, defects in the video-detector tube will cause the same symptoms as defects in the video-IF amplifier stages. In the majority of receivers, a semiconductor diode (Fig. 2) instead of a tube is used as the video detector.

A dead video-amplifier tube will result in no picture and no snow, while a tube with low emission will cause a dim picture that lacks contrast. Other symptoms which may be caused by defects in the video-amplifier tube are excessive

Fig. 11. Smeared picture.

Fig. 12. Poor focus.

Fig. 13. Hum bar in picture.

contrast (Fig. 10), negative picture (Figs. 8 and 9), smeared picture (Fig. 11), or loss of picture detail.

A defective picture tube can also produce most of the symptoms which have been described for the other sections of the video channel—no picture (with no snow), weak picture (with dim raster), smeared picture, and negative picture. In addition, a defective picture tube can result in poor focus (Fig. 12), or even in complete loss of raster (dark screen).

Fig. 14. Picture pulling.

Fig. 15. Effect of heater-to-cathode leakage in local oscillator.

Hum in the picture is a common symptom and can be caused by heater-to-cathode leakage in any tube carrying the picture signal. Fig. 13 shows one example. Depending on the amount of leakage and the signal level at the point of leakage, the hum bar will be lighter or darker. In any event, a single dark and light bar across the picture signifies heater-to-cathode leakage in the video channel. Hum may also appear as picture pulling, as seen in Fig. 14.

Even though the local-oscillator tube does not carry the picture signal, heater-to-cathode leakage will cause the picture symptom in Fig. 15—a horizontal band that becomes brighter as the leakage increases. At excessively high leakage, the oscillator stops and the screen has a raster only, with no evidence of hum voltage. In receivers having four IF-amplifier tubes, noise (snow) is prominent; but with three IF tubes, the noise is less prominent; and if only two IF tubes are used, it is practically invisible.

Heater-to-cathode leakage in the picture tube does not produce hum symptoms in the picture because the AC level is lower than the video-signal level. However, in most receivers such leakage disturbs the cathode bias, causing an abnormally bright picture (Fig. 16) over which the brightness control has little or no effect.

Fig. 16. Effect of heater-to-cathode leakage in picture tube.

Sound Troubles

Failure of a tube in the sound channel (sound-IF amplifier, audio-detector, AF amplifier, and audio-output) will result in no sound output. The picture will generally be unaffected; however, there are exceptions which will be discussed later under interaction. Usually, if the picture and/or sync are affected along with the sound, look for trouble in the video channel prior to the sound take-off point.

A tube with low emission in the sound channel will result in reduced volume in the sound output. A gassy tube or one with heater-to-cathode leakage may produce a garbled or distorted output and, if severe enough, will completely blank out the audio, replacing it with 60-cycle hum. A defective detector tube can cause sync buzz; that is, the desired audio signal is accompanied by a buzzing sound. To distinguish between hum and buzz, remember that hum is a soft 60-cycle tone, and that buzz is a harsh, rasping 60-cycle tone.

AGC Troubles

Since, as explained previously, the AGC tube controls the gain of the video-IF amplifiers, a defective AGC tube will produce some of the same symptoms given for defective tubes in the video channel. The most common symptoms caused by defective AGC tubes are no picture or sound

(Fig. 6), negative picture (Figs. 8 and 9), excessive contrast (Fig. 10) usually accompanied by sync buzz in the sound, and picture pulling (Fig. 14).

Sync Troubles

Tube defects in the sync and noise-canceller sections of the receiver will disrupt the normal separation of the video signal from the sync pulses. In addition, noise pulses will

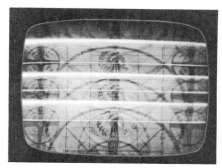

Fig. 17. Loss of vertical sync.

Fig. 18. Loss of horizontal sync.

not be removed from the signal, and coupling of the vertical- and horizontal-sync pulses to their respective oscillators will be disrupted. Thus, depending on the extent of the defect and on the circuit, a faulty tube in the sync section may cause loss of vertical sync, resulting in the picture rolling up or down (Fig. 17); or loss of horizontal sync, where the picture breaks up into a series of diagonal horizontal bars (Fig. 18). At other times, both the vertical and the horizontal sync may be lost, as shown in Fig. 19. Sometimes a defect in the sync stage will produce vertical jitter; that is, the picture will bounce up and down, or roll partway

Fig. 19. Loss of both vertical and horizontal sync.

Fig. 20. Horizontal bending.

Fig. 21. Vertical-sync pulse absent but oscillator on frequency.

and then return. Likewise, a defective sync tube can produce horizontal bending or pulling in the picture, as shown in Fig. 20.

Since loss of horizontal or vertical sync can also be caused by a defective oscillator tube, a quick check known as *freewheeling* can sometimes be made to determine whether the defect is in the sync or the oscillator section. Try critically adjusting the hold controls to bring the oscillator back to

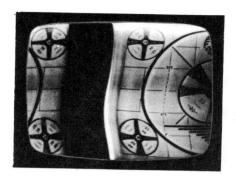

Fig. 22. Horizontal-sync pulse absent but oscillator on frequency.

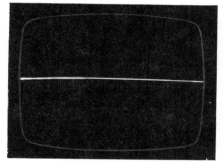

Fig. 23. Loss of vertical sweep.

the correct frequency. If the picture momentarily returns to sync, the oscillator is probably all right. However, even though the oscillator frequency is correct, it may not be in phase with the transmitted signal and the resulting picture will look like Fig. 21 or 22.

Vertical Troubles

Complete failure of either the vertical-oscillator or -output tube will result in no vertical sweep, seen as a thin horizontal line in the picture (Fig. 23). Should the oscillator or output tube have low emission, the height will be reduced. As mentioned under sync troubles, a defective oscillator tube can also cause loss of vertical sync and the resulting picture roll. However, if the vertical oscillator is at fault and you try "freewheeling" the picture, it will lock in phase if you can bring the oscillator to the correct frequency with the hold control. Often, the picture can be locked in sync at only one extreme of the hold control when the oscillator tube is at fault.

Fig. 24. Insufficient height and poor vertical linearity.

Fig. 25. Loss of horizontal sync, and insufficient width.

Partial failure of the vertical-output tube usually results in insufficient height and poor linearity, as shown in Fig. 24; that is, the horizontal lines in the raster are unevenly spaced (either cramped or stretched out) at the top or bottom.

Horizontal and High-Voltage Troubles

Failure of the horizontal-AFC tube will usually result in loss of horizontal sync, or in improper horizontal phasing. These symptoms were shown in Figs. 18 and 22. [Note: In many chassis, semiconductor diodes (Fig. 2) are used for the horizontal-AFC function.] Complete failure of the horizontal-oscillator, horizontal-output, damper, and high-voltage rectifier tubes will cause loss of raster (a black screen with no light); likewise, low emission in any of these tubes can cause insufficient width. If insufficient width is accompanied by loss of horizontal sync as shown in Fig. 25, the oscillator is probably at fault.

Defects in the horizontal-oscillator tube will also cause loss of horizontal sync (Fig. 18), improper phasing (Fig. 22), or horizontal pulling (Fig. 20).

Fig. 26. Barkhausen oscillations.

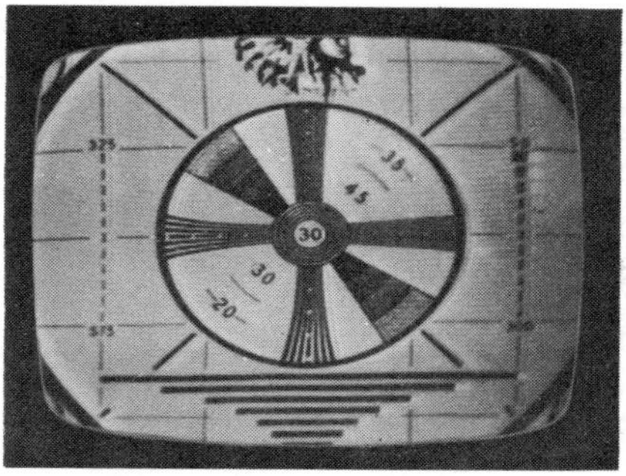

Fig. 27. Blooming, due to low high voltage.

Barkhausen lines, seen at the left edge of the raster in Fig. 26, are caused by a defect in the horizontal-output stage. They can sometimes be eliminated by replacing the horizontal-output tube.

As mentioned previously, a weak horizontal-output tube will cause the picture to pull in from the sides. In addition, the picture is often dim because the horizontal-output tube drives the damper, which in turn feeds DC plate and screen voltages to various other tubes. The picture may also bloom because the horizontal-output tube drives the high-voltage rectifier tube. Blooming (Fig. 27) is the phenomenon in which the raster expands and defocuses as the brightness control is advanced; it is caused by enlargement of the spot due to halation. For proper focus, it is necessary to reduce the brightness level below normal and the screen may not be filled. Also, the raster may disappear as the brightness control is advanced near its maximum setting. This condition is caused by insufficient high voltage—either the pulse from the horizontal-output tube is too low or the high-voltage rectifier is weak or dead.

Fig. 28. Poor horizontal linearity on right side.

Fig. 29. Poor horizontal linearity on left side.

A defective horizontal-output tube can also cause poor horizontal linearity, as shown in Fig. 28. Notice that the right side of the raster is compressed. If the left side is compressed (Fig. 29), the damper tube is weak.

Low-Voltage Troubles

Total failure of the low-voltage rectifier tube(s) (or semiconductor) will result in complete loss of raster and sound. (If the trouble is in the horizontal circuits, the sound will usually not be affected.) Partial failure can result in insufficient width and height, poor picture quality, and weak sound. Usually, however, the reduced width and height are the most easily recognizable symptoms.

VARIATIONS IN TROUBLE SYMPTOMS

So far we have pinpointed various troubles to their respective section in a TV receiver. Sometimes, however, a given defect will produce symptoms different from those explained earlier. The reasons for this—interaction between functions, variations in layout of a receiver or in the filaments, etc.—will now be discussed.

Interacting Functions

As mentioned previously, the damper tube supplies a boost voltage to the picture tube, sound detector, sync-separator, vertical-output, horizontal-oscillator, and horizontal-output tubes (although they all may not use it). Thus, a weak damper tube can reflect trouble symptoms into any of these other tubes—the degree and number depending on the tube types and on the individual circuit.

In some receivers, DC voltage is applied from the damper section to the horizontal-output tube only. Hence, with limited interaction, interpretation of picture and sound symptoms is greatly simplified. However, damper-tube interaction is not the only complication to look out for. In numerous receivers you will find interaction between the audio-output tube and other functions. Here, plate and screen DC voltages may be fed from the audio-output section to the mixer-oscillator, IF and video-amplifiers, sync-phase inverter, sync-separator, horizontal-output, and AGC-keyer tubes. Consequently, a weak or dead audio-output tube will cause prominent picture trouble symptoms.

Dual-section tubes also cause interacting symptoms in many receivers. For instance, a receiver may employ two 6BA8 triode-pentode tubes, half of the first tube acting as a video-IF amplifier and the other half as the horizontal oscillator. Likewise, the second tube may be the video amplifier and sync separator. If the cathode in the first 6BA8 has low emission, both the IF-amplifier and the horizontal-oscillator functions will be adversely affected; and if the cathode in the second tube has low emission, this will affect operation of both the video-amplifier and the sync-separator functions. The reason is that a 6BA8 tube has a single heater common to both sections.

Some tubes such as the 12AU7 have a separate heater for each section. In parallel-heater systems (discussed in the next section) the heater often weakens or burns out in one section leaving the other section unaffected. Thus, a burned-out heater in one section of the 12AU7 does not necessarily impair operation of the other section. So, picture and sound symptoms must often be interpreted with respect to the heater structure of a given tube as well.

Parallel Heaters

In the series-heater strings depicted in Fig. 30A, all tubes go "dead" when a heater burns out, because all heaters are connected in series. In parallel-heater arrangements, shown in Fig. 30B, the remaining tubes continue to operate when a heater burns out. Hence, parallel-heater systems are easier to service because any unlit tubes can easily be checked and replaced.

Heater-to-cathode shorts sometimes cause unexpected reactions in series-string receivers. If the cathode of the stage is grounded and a heater-to-cathode short develops, no filament voltage will be applied to the tubes following the shorted one, making them inoperative, and the entire 117-volt applied voltage will be dropped across the remaining tubes. Should this occur near the end of the string—for example, in the horizontal AFC-oscillator tube in Fig. 30A—the picture, RF amp, and oscillator-mixer tubes would be inoperative and, although it might not be noticeable, the remaining tubes would glow more brightly than normal. The symptom will be no raster and no sound; but the cause is the horizontal AFC-oscillator tube—instead of the low-voltage rectifier as might normally be suspected.

If a heater-to-cathode short should develop near the input of the filament string—say, the sync amplifier in Fig. 30A—the filament voltage would be removed from all of the remaining tubes as before. However, the entire 117 volts will now be dropped across only four tubes, so the filament of one or more will probably burn out because of this excessive drop. Thus, when replacing a tube with a burned-out filament in a series string, make sure another shorted tube or component is not the real reason for the defect. Otherwise, the new tube will blow, too.

Functional Arrangement of Receivers

The functional arrangement of receivers will vary slightly from manufacturer to manufacturer, and will have an effect

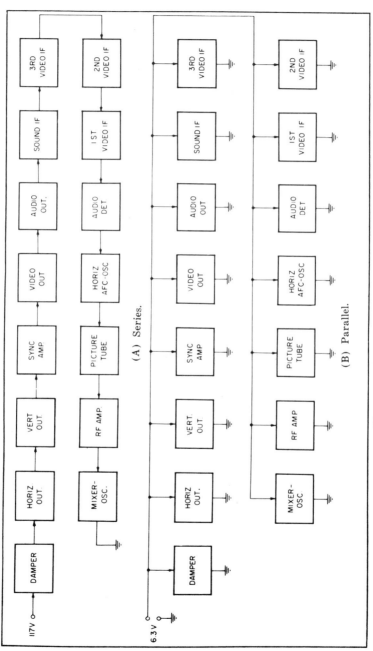

Fig. 30. Series and parallel filaments.

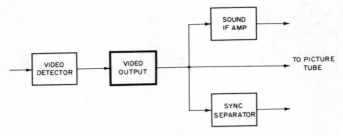

(A) Both following video output.

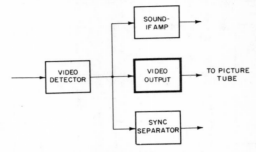

(B) Both preceding video output.

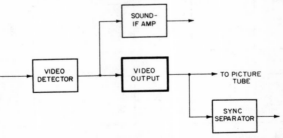

(C) Sound take-off preceding and sync take-off following video output.

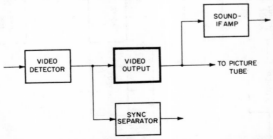

(D) Sound take-off following and sync take-off preceding video output.

Fig. 31. Four methods of sync and sound take-offs.

on the symptoms produced should a given tube fail. For example, Fig. 31 shows four arrangements for the sound and sync take-offs. Failure of a given tube—in this example, the video output—will produce varied symptoms. In Fig. 31A, both take-offs follow the video output; therefore, there will be no sound or picture, and the vertical-retrace lines will move up and down on the raster as shown in Fig. 32A. In

(A) Unlocked. (B) Locked.
Fig. 32.Unlocked and locked vertical-retrace lines.

Fig. 31B, both take-offs precede the video amplifier, so sound will be present and the vertical-retrace lines will be stationary as shown in Fig. 32B, indicating that sync pulses are reaching the vertical oscillator. In Fig. 31C, sound will be present and the vertical-retrace lines will be unlocked; and in Fig. 31D the sound will be missing but the vertical-retrace lines will be locked.

Thus, before properly evaluating the symptoms, you should determine several points about the receiver. Study the schematic first, to see (1) where the various take-offs are, (2) what stages use the boost voltage, (3) if some of the tubes obtain B+ from the audio-output or some other tube, and (4) if the filaments are in series or parallel. Keep these points in mind, and the symptom will tell you much more about the possible causes.

UHF TUNER

All modern TV receivers have UHF tuners. If picture and sound reproduction are normal on VHF channels, but reception is weak, distorted, or absent on UHF channels, the tube in the UHF tuner is a likely culprit. Some UHF tuners use transistors instead of tubes. Transistor replacement is beyond the scope of this book.

HYBRID RECEIVERS

There is a trend in modern TV receivers toward hybrid arrangements. This refers to the use of both transistors and tubes. The following arrangements are typical:

1. Transistors used in first and second sound-IF stages and in the first AF stages; tubes used in all other stages.
2. Transistor used in UHF tuner; tubes used in all other sections.
3. Transistors used in UHF tuner and in AGC circuit; tubes in all other sections.
4. Transistors used in UHF tuner, in first and second sound-IF stages, and in first AF stage; tubes used in all other sections.
5. Transistors used in UHF tuner, and in first, second, third, and fourth IF stages; tubes used in all other sections.
6. Transistors used in UHF tuner, and in noise-gate circuit; tubes used in all other sections.

This book applies to trouble symptoms and replacement of tubes in hybrid receivers, but does not cover replacement of transistors. Reference should be made to specialized TV trouble-shooting books for coverage of transistor defects and replacement.

COLOR RECEIVERS

You should note that all of the TV tube symptoms and troubles explained in this book apply to both black-and-white TV receivers and to the *black-and-white section* of a color-TV receiver. In other words, a color-TV receiver consists basically of a black-and-white section and a chroma section, as shown in Fig. 33.

In general, tube troubles in the chroma section of a color-TV receiver affect color reproduction only. If reception of black-and-white programs is weak, distorted, or absent, and if symptoms of sound trouble may be present, it is most probable that there is a defective tube in the black-and-white section of the color receiver. Troubles in the chroma section

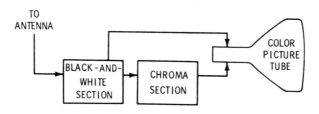

Fig. 33. Simplified block diagram of a color receiver.

are discussed in more specialized TV troubleshooting books and are beyond the scope of this guidebook.

OTHER HELPFUL HINTS

If visual examination reveals a tube is not lit, it probably is defective. Sometimes a tube may not appear lit at first glance, but closer examination will reveal that it is. As mentioned before, one open filament in a series string will cause all tubes in a receiver to be dark.

CAUTION: Do not touch the horizontal-output tube or high-voltage rectifier while the set is plugged in. The high-voltage rectifier will not glow or feel warm, even when operating properly.

If a tube glows exceedingly brightly or its plate is red-hot, carefully substitute another tube. If the new tube does the same, quickly remove the power and look for a defect in the circuitry.

A blue glow can often be seen inside the glass envelope of tubes handling appreciable power, such as low-voltage rectifier and horizontal-output tubes. This surface glow, called fluorescence, is a normal condition—it does not indicate a faulty tube. On the other hand, if the blue glow is inside the tube, around the elements, the tube is gassy. In an audio-output tube, for example, traces of gas will cause distorted sound or none at all. Voltage amplifiers (such as IF tubes) handle comparatively little power and seldom show a blue glow if gassy. However, the picture will be distorted or absent, but the sound will often remain unchanged.

Beginners are sometimes puzzled by the problems created by a gassy tube. Most tube testers have a gas test; however, a gas test applies only to tubes with grids, and does not apply to diodes. Some tube testers do not provide a gas test. Nevertheless, it is easy to tell with any tube tester whether any tube has quite a bit of gas. Consider a diode tube. If the tube is quite gassy, there will be a greatly abnormal current flow from cathode to plate—in turn, the pointer on the meter scale will rapidly move off to the right. The same observation holds for triodes, tetrodes, and pentodes—if the tube is quite gassy, the pointer will be pinned at the right-hand end, off scale. You will also see a blue glow around the elements inside the tube, just as when it is operating in a receiver.

In the case of high-voltage rectifier tubes, gas frequently causes internal arcing, which produces streaks in the picture, and is often accompanied with loss of sync. In color-TV receivers, sync lock may not be lost, but the gas in a high-voltage rectifier tube can cause circuit disturbances that make the picture wave erratically. In any case, first thing to do is to examine the tubes carefully for evidence of a blue glow around and inside the plate.

In older receivers having a large number of tubes, blue flashes can sometimes be seen in the low-voltage rectifier tube while the receiver is warming up. This is an arc-over or breakdown flash, which tends to shorten tube life. Connection of a current-limiting device in series with the 117-volt lead to the power supply will eliminate this difficulty, unless there is trouble in the circuit. The current limiter restricts the AC current flow into the receiver until the tubes have warmed up.

TUBE TESTING

Since no tube tester can be 100% perfect, the user should be aware of the limitations which may occasionally pass a "bad" tube as "good." One of the most common tube defects is interelectrode leakage—particularly heater-to-cathode leakage. For this reason, all tube testers provide for leakage and short-circuit defects. One of the limitations in ordinary tube testers is the amount of leakage which is permitted before a "bad" indication is provided. An extremely sensitive leakage test would result in the rejection of many tubes that operate satisfactorily in the majority of TV receiver circuits. On the other hand, there are a minority of receiver circuits that do not operate properly unless the tubes have very little interelectrode leakage. In such cases, tubes that are passed as "good" by an ordinary tube tester might not operate properly in such sensitive circuits. You might have to select a suitable tube from among several "good" tubes.

Ordinary tube testers commonly provide an emission test to indicate the merit of a tube. An emission test is very useful, but it also has certain limitations. Sometimes a tube will develop a cathode "hot spot." This is not a common occurrence, but it does happen. In such case, the tube is likely to give poor performance in a receiver, although an emission tester passes the tube as "good." There is another limitation to be recognized; it concerns the reduction in emission which is permitted before a "bad" indication is provided. Some TV receiver circuits require more cathode emission than other circuits. Hence, it is possible that a tube which has low emission, and tests "bad," may operate satisfactorily in certain circuits.

The simpler types of emission testers do not indicate whether a screen grid, for example, might be open circuited. In such case, the tester might pass the tube as "good" although it would fail to function satisfactorily in a receiver circuit. This type of defect is comparatively rare, although it does occur. The more elaborate emission testers provide a series of switching tests that permit indication of an open screen grid or other electrode in a tube.

There is another limitation encountered in the simpler types of emission testers which concerns heavy-current tubes. Some of the modern output tubes normally operate with comparatively large current flow from cathode to plate. Unless the tube tester can provide an appropriate amount of current, an output tube might be passed as "good" although its performance is somewhat weak in a receiver circuit. The advent of wide-angle large-screen receivers has placed increasing demands on the current capability of emission-type tube testers.

In summary, ordinary tube testers, particularly those of the simpler emission types, are very useful and their indication can be relied on in the great majority of cases. However, there will always be a chance that a certain tube, used in a particular receiver circuit, might be passed as "good" although its performance is unsatisfactory. Conversely, there is always a chance that a tube rejected by a simple tube tester as "bad," might perform satisfactorily in certain TV receiver circuits.

GENERAL CHART OF SYMPTOMS

The chart of symptoms on the following pages gives a brief summary of various picture and sound symptoms resulting from tubes having low or no emission and those which are gassy. The common symptoms are noted without reference to less common interacting functions. Thus, although the chart serves as a guide, interacting functions must sometimes be taken into account as well.

For an explanation of the notes, see the end of the chart.

GENERAL CHART OF SYMPTOMS

TUBE	DEFECT	SYMPTOM	
		Picture	Sound
RF amplifier.	Low emission.	Weak, snowy.[1]	Normal.
	No emission.	Weak or no picture; snow.[2]	Often weak and noisy.
	Gassy.	Muddy, filled up; raster may be streaked on vacant channels.[3]	Normal.
	Heater-to-cathode leakage.	Hum bar in picture; modulated noise bar on vacant channel.[4]	Normal, unless hum is very high.[5]
Mixer.	Low emission.	Weak, snowy.[1]	Normal.
	No emission.	Perceptible snow in raster; no picture.[1]	No sound; noise from speaker when volume control is turned up.
	Gassy.	Weak or no picture; snow.	Often weak and noisy, or no sound.
	Heater-to-cathode leakage.	Hum bar in picture; modulated noise bar on vacant channel.[4]	Normal, unless hum is very high.[5]

Oscillator.	Low emission.	Weak.	Normal.
	No emission.	Prominent snow in raster; no picture.[1]	No sound; noise from speaker when volume control is turned up.
	Gassy.	Weak or no picture; snow.	Often no sound; noise from speaker when volume control is turned up.
	Heater-to-cathode leakage.	Hum bar.[6]	Normal, unless hum is very high.[5]
First IF.	Low emission.	Weak.	Normal.
	No emission.	Weak or no picture.	Normal or noisy.
	Gassy.	Muddy, filled up.	Normal.
	Heater-to-cathode leakage.	Hum bar in picture; modulated noise bar on vacant channel.[4]	Normal, unless hum voltage is very high.[5]
Second, third, or fourth IF.	(Same as for the first IF tube, except that feedthrough takes place in the last IF stage, even when the signal is strong. No emission results in a blank raster.)		

39

GENERAL CHART OF SYMPTOMS—(Cont'd)

TUBE	DEFECT	SYMPTOM	
		Picture	Sound
Video detector.	Low emission (or low front-to-back ratio if crystal diode is used).	Weak.	Normal or noisy.
	No emission (or open or shorted crystal diode).	No picture (raster only).	No sound.
	Gassy.	Weak or no picture.	Noisy or no sound.
	Heater-to-cathode leakage.	Hum bar in picture.	Normal, unless hum is excessive.
Video amplifier.	Low emission.	Low contrast.	Normal.
	No emission.	Raster only.[7]	Normal or no sound.[8]
	Gassy.	Muddy, filled up; sync often unstable.	Normal or noisy sound.[9]
	Heater-to-cathode leakage.	Hum bar in picture.	Normal, or buzz interference if hum level is excessive.
Picture tube.	Low emission.	Dim.	Normal.

40

No emission.	Dark Screen.	Normal.
Gassy.	Dim, "silvery."	Normal, or buzz in sound if excessively gassy.
Heater-to-cathode leakage.	Little or no control of brightness.	Normal, or buzz in sound if brightness is excessive.
AGC keyer.		
Low emission.	Picture muddy, filled up except on weak channels; stronger channels may have negative reversal and sync buzz, or no picture.	Harsh 60-cycle rasp (sync buzz) may be present. (No sound when picture is missing.)
No emission.	Negative reversal and sync buzz, or no picture.	Harsh rasp. (No sound when picture is missing.)
Gassy.	(Same as for low emission.)	
Heater-to-cathode leakage.	Hum bar in picture.	Usually normal, but may have 60-cycle interference in extreme cases.

GENERAL CHART OF SYMPTOMS — (Cont'd)

TUBE	DEFECT	SYMPTOM	
		Picture	Sound
AGC clamper.	Low emission.	No picture on weak channels; normal reception on stronger channels.	Noisy or no sound on weak channels; normal sound on stronger channels.
	No emission.	Weak or snowy picture on stronger channels; no picture on weak channels.	Noisy or no sound.
	Heater-to-cathode leakage.	Hum bar in picture unless signal is quite strong.	Usually normal, but may have 60-cycle interference unless signal is strong.
Noise canceller.	Low emission.	Unstable sync.	Normal.
	No emission.	No sync lock.	Normal.
	Gassy.	Unstable sync.	Normal.
	Heater-to-cathode leakage.	Bent picture; sync often unstable.	Normal.
Sync separator.	Low emission.	Unstable horizontal and	Normal.

42

Stage	Condition	Symptom	
	No emission.	vertical sync.	Normal.
	Gassy.	Horizontal sync very critical or absent; no vertical sync.	Normal.
	Heater-to-cathode leakage.	Unstable or no horizontal and vertical sync.	Normal.
		Bent picture or loss of sync.	Normal.
Sync-phase inverter.		(Same as for sync separator.)	
Vertical oscillator.	Low emission.	Picture lacks height and often rolls vertically.	Normal.
	No emission.	Bright horizontal line across center of screen.	Normal.
	Gassy.	Picture lacks height, rolls, and may collapse intermittently.	Normal.
	Heater-to-cathode leakage.	Picture lacks height, has vertical nonlinearity, and often locks out of phase (or rolls and jumps).	Normal.

GENERAL CHART OF SYMPTOMS—(Cont'd)

TUBE	DEFECT	SYMPTOM	
		Picture	Sound
Vertical output.	Low emission.	Bottom is compressed and often folded over.	Normal.
	No emission.	Bright horizontal line across center of screen.	Normal.
	Gassy.[10]	Bottom compressed; tube may arc intermittently and cause picture to collapse.	Normal, except for clicks on intermittent collapse.
	Heater-to-cathode leakage.	Top expanded off the screen.	Normal.
Horizontal AFC.	Low emission (or low front-to-back ratio of crystal diodes).	Unstable horizontal sync; normal vertical sync.	Normal.
	No emission (open or shorted crys. diodes).	Horizontal sync very critical or absent, vertical sync normal.	Normal.
	Gassy.	Unstable or no horizontal sync; normal vertical	Normal.

		sync.	
Horizontal oscillator.	Heater-to-cathode leakage.	Bent picture or no horizontal sync; normal vertical sync.	Normal.
	Low emission.	Narrow and nonlinear picture; horizontal sync often unstable.	Normal.
	No emission.	Dark screen.	Normal.[11]
	Gassy.	(Same as for low or no emission.)	
Horizontal output.	Heater-to-cathode leakage.	Narrow, bent; horizontal sync often unstable.	Normal.
	Low emission.	Narrow, nonlinear; blooming in severe cases.	Normal.
	No emission.	Dark screen.	Normal.[11]
	Gassy.[10]	(Same as for low or no emission, except the tube runs very hot.)	
	Heater-to-cathode leakage.	Excessive leakage causes a wide range of symptoms, or no reception in series-heater layouts.	Sound weak and has buzz interference, or is absent (symptoms vary widely).

45

GENERAL CHART OF SYMPTOMS—(Cont'd)

TUBE	DEFECT	SYMPTOM	
		Picture	Sound
Damper.	Low emission.	Narrow, nonlinear; and blooms in severe cases.	Normal.
	No emission.	Dark screen.	Normal.[11]
	Gassy.[10]	(Same as for low or no emission.)	
	Heater-to-cathode leakage.	Horizontal streaks or intermittent collapse.	Popping noises.
High-voltage rectifier.	Low emission.	Dim, and blooms when brightness control is turned up.	Normal.
	No emission.	Dark screen.	Normal.
	Gassy.	(Same as for low or no emission, sometimes accompanied by horizontal streaks and popping noises.)	
Power-supply rectifier.	Low emission (low front-to-back ratio if semi-conductor rectifier is used).	Narrow, often nonlinear, and lacks height.	Usually normal, sometimes weak.

	No emission.	Dark screen.	No sound.
	Gassy.	Narrow, often nonlinear, and lacks height. If tube is very gassy, screen will be dark.	Normal (no sound when screen is dark).

1 High-performance receivers with four IF stages display snow prominently; receivers with three IF stages display moderate snow; and those with only two IF stages display practically none. This assumes the use of a rabbit-ear antenna, or an external antenna which picks up a negligible amount of man-made noise.
2 Sensitive receivers will display a weak picture, even if the RF amplifier is dead, because of the feedthrough signal. Even a simplified receiver will display a weak picture if the incoming signal is very strong.
3 Appreciable gas causes the grid to operate at a very low or positive bias; the tuned circuits occasionally become unstable and have regeneration or weak oscillation, displayed as long horizontal noise streaks in the picture.
4 Snow (noise) voltages are like signal voltages, and if the receiver sensitivity is high, heater-to-cathode leakage will cause a modulated noise bar to be displayed in the raster. On the other hand, if the receiver sensitivity is low, the raster will remain clear on vacant channels.
5 When the hum voltage is high enough to cut off the tube over part of the cycle, the sound signal will be "killed" at a 60-cycle rate, causing a hum or buzz in the sound.
6 Horizontal sync sometimes impaired (picture "pulls").
7 Vertical-retrace lines are locked-in if sync take-off precedes the video amplifier. Retrace lines will run at random if sync is taken off at the video-amplifier output.
8 Sound will be normal if taken off before video amplifier, absent if taken off at video-amplifier output.
9 Sound will be normal if taken off before video amplifier. When sound is taken from video-amplifier output, gas current flow can limit video signal and cause sync buzz.
10 A tube that is sufficiently gassy to cause noticeable picture symptoms can be spotted by the telltale blue ion glow *inside* the tube, between and around the elements. Do not confuse it with the blue fluorescent glow on the inner glass envelope of many normally operating tubes.
11 In receivers using a boost B+ voltage in the sound section, the sound may be weak, distorted, or absent.

PICTURE-TUBE DISPLAYS

In this section, various symptoms traceable to tube faults will be given, and the tubes which can cause these symptoms will be pointed out. The troubles can be categorized as follows:

>Picture
>Sync
>Size and Linearity
>Raster
>Sound

Each symptom is placed in the most prominent category: For example, if a defect in a signal circuit causes loss of sync and also affects the picture somewhat, the symptom will be displayed under sync trouble because it is the more noticeable. Likewise, a defect may show up in the picture or sync, and also in the sound. In this case the symptom will be given under the appropriate picture-tube symptom, and its effect on the sound will also be noted.

A discussion of the troubles in each category precedes the picture-tube symptoms illustrated and explained on the following pages. Although the most usual situations are covered, remember that TV receivers differ widely in layout. Hence, always study the circuit diagram of the receiver first so you can form a mental block diagram of its layout, as explained in Section 1. Then you will be able to analyze the picture and sound symptoms with maximum exactness.

Although you should always check tubes first in localizing any symptoms, keep in mind that the same symptoms are sometimes caused by defective circuitry. There is no simple method of determining whether a negative picture, for example, is caused by a faulty AGC tube or by a leaky delay capacitor, except by substituting a new tube. When this doesn't work, instruments such as the VOM, VTVM, and oscilloscope will be necessary. These are explained in more specialized TV troubleshooting books and are beyond the scope of this book.

PICTURE TROUBLES

Picture troubles are those which primarily affect the presence (or absence) and the quality of the picture. There will be a raster but no picture, or if present the picture may not be reproduced properly. Such troubles include no picture or one that is snowy, weak, smeared, or negative; poor focus; improper brightness; hum; and various others. The effect on the sound in a typical circuit is also given; however, since the sound takeoff points will vary, the effect on the sound may also vary. Hence, all sound symptoms should be evaluated with respect to the particular receiver in question.

Trouble   *Page*

No Picture; Raster Normal; No Snow; Weak or Noisy Sound 53
No Picture; Raster Normal; Snow Level High; No Sound 53
No Picture; Raster Normal; Snow Level Medium; No Sound 54
No Picture; Raster Normal; Snow Level High and Streaked;
    No Sound 54
No Picture; Raster Normal; Vertical-Retrace Lines Locked-in;
    No Snow; Sound Normal 55
No Picture; Raster Normal; Vertical-Retrace Lines Not
    Locked-in; No Snow; Sound Normal 55
No Picture; Strong Hum Bar in Raster; 60-Cycle Hum in Sound .. 56
No Picture; Strong Hum Bar with Herringbone in Raster;
    60-Cycle Hum in Sound 56
Hum Bar with Snow on Unused Channels 57
Picture Weak and Silvery; Raster Dim; Sound Normal 57
Picture Weak and Bent; Raster Normal; Sound Normal 58
Picture Contrast Poor; Screen Brightness Normal 58
Picture Wavy; Brightness Excessive, with Little or No Control 59
Picture Muddy and Filled Up; Sync Buzz in Sound 59
Negative Picture; Sync May Be Unstable; Sync Buzz in Sound 60
Smeared Picture 60
Trailing Reversal 61
Poor Focus; Sync and Sound Normal 61

continued on following page

51

Trouble	Page
Retrace Lines Prominent and Jittery; Picture Often Drifts Up and Down Slightly	62
Picture Bent, Washed Out, and Exhibits a Bright Hum Bar	62
Prominent Hum Bar in Picture, Sync Lock Normal	63
Hum in Both Picture and Sound	63
"Pie-Crust" Pattern in Picture	64

No Picture; Raster Normal; No Snow; Weak or Noisy Sound

This symptom is often caused by a dead tube in the last IF stage. No picture signal feeds through the dead tube, but the sound signal often does (although weak or noisy). The volume control must be turned higher than normal, and noise is heard because the audio-detector operation is subnormal. In some receivers a defective audio-output tube can also cause this symptom.

This screen symptom points to a faulty local-oscillator tube because most snow in the raster is generated in the mixer tube. (If the RF-amplifier tube were dead, we would usually see a weak picture on the stronger channels.) No sound confirms a dead oscillator tube.

No Picture; Raster Normal; Snow Level High; No Sound

No Picture; Raster Normal; Snow Level Medium; No Sound

A large reduction in the snow level indicates a dead mixer tube. The amount of snow will depend on the number of IF's —the more IF's, the more snow. (Receivers with only two IF stages usually do not display snow when the mixer tube is dead.) No sound also confirms a dead mixer tube.

The high snow level and the fact that no picture is displayed (even on strong channels) points to a dead local-oscillator tube, as previously noted. However, *streaked* noise indicates IF regeneration, often due to substitution of another tube type with excessively high g_m.

No Picture; Raster Normal; Snow Level High and Streaked; No Sound

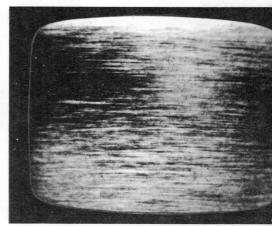

No Picture; Raster Normal; Vertical-Retrace Lines Locked-in; No Snow; Sound Normal

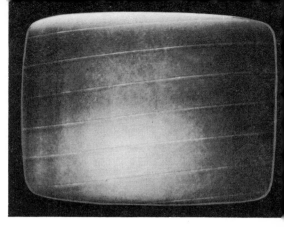

This symptom indicates a dead video-output tube in receivers in which both the sync and the sound are taken off before the video-output tube. (If the sound is taken off after the video-output tube, there will be no sound.) If DC coupling is employed between the video-output tube and the picture-tube cathode, the picture-tube bias will be raised, and the raster will be dim.

Another indication of a dead video-output tube. This symptom applies to receivers in which the sync take-off is after the video-output tube. The vertical-retrace lines move up and down the screen at random because the sync action is lost. As before, if the sound is taken off after the video-output tube, there will be no sound; and if DC coupling is employed, the raster may be dim.

No Picture; Raster Normal; Vertical-Retrace Lines Not Locked-in; No Snow; Sound Normal

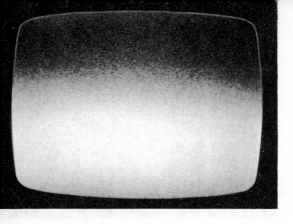

No Picture; Strong Hum Bar in Raster; 60-Cycle Hum in Sound

This symptom illustrates a "paralyzed" RF-amplifier tube: No picture is displayed, even on strong channels. Heater-to-cathode leakage in front-ends, such as cascode units, causes the plate voltage from the RF-amplifier tube to bleed to ground, and also injects hum voltage into the signal circuit.

Here is another example of heater-to-cathode leakage in a cascode RF-amplifier tube. When this happens, the signal circuits are thrown out of correct alignment and spurious branch circuits are set up. In some receivers the RF-amplifier circuit will oscillate, causing a herringbone pattern in the hum as shown.

No Picture; Strong Hum Bar with Herringbone in Raster; 60-Cycle Hum in Sound

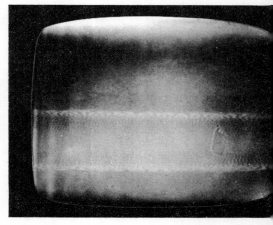

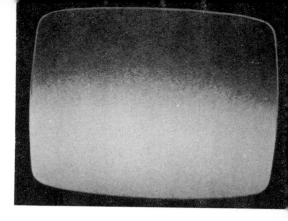

Hum Bar with Snow On Unused Channels

When there is heater-to-cathode leakage in a front-end or IF tube, we might suppose there would be no screen symptom on *vacant* channels. However, a hum-modulated noise pattern may be present in a sensitive receiver. In other words, except for being random in structure, noise voltages act like video-signal voltages.

This symptom can be caused by a gassy picture tube, particularly if the picture has a shimmering or silvery cast.

Picture Weak and Silvery; Raster Dim; Sound Normal

Picture Weak and Bent;
Raster Normal;
Sound Normal

This is another indication of a gassy picture tube. If the sync is taken off at the picture-tube input, the picture will often appear bent as shown here.

Poor contrast is usually caused by a weak or dead tube in the signal channel. Check all tubes progressively, from the RF amplifier to the video amplifier. Usually the sound signal will not be noticeably reduced, but it may be.

Picture Contrast Poor;
Screen Brightness Normal

Picture Wavy; Brightness Excessive, with Little or No Control

In most cases, heater-to-cathode leakage in the picture tube causes loss of bias and hence the excessive brightness. No hum bar appears because the signal level is comparatively high, although the picture may be expanded.

This symptom is often caused by a gassy AGC tube or one with low emission. Also look for a gassy video-IF amplifier or video-amplifier tube.

Picture Muddy and Filled Up; Sync Buzz in Sound

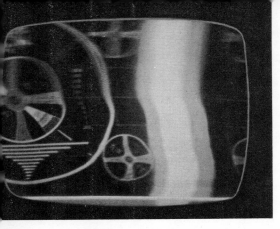

Negative Picture; Sync
May Be Unstable;
Sync Buzz in Sound

This is a more pronounced combination of the previous symptom, involving the same tubes. If the tubes are in good condition, the trouble is in the circuitry associated with these stages.

A gassy video-amplifier tube, or one with an open screen, will result in a smeared picture. If this tube is in good condition, check the video-detector tube (or semiconductor diode).

Smeared Picture

Trailing Reversal

Trailing reversal is caused by incorrect bandwidth in a signal channel. A shield may have been left off a high-frequency tube, or a tube with excessively high g_m may have been substituted for an IF amplifier.

Poor focus, when contrast and brightness are normal, can be caused by an erroded first anode in the picture tube. This trouble is due to a misadjusted ion trap, and is most likely to occur in large-screen tubes operating at a high second-anode voltage.

Poor Focus; Sync and Sound Normal

Retrace Lines Prominent and Jittery; Picture Often Drifts Up and Down Slightly

Heater-to-cathode leakage in the keyed-AGC tube disturbs the bias and causes the washed-out picture. The hum bar is narrow and occurs during the vertical-blanking interval only.

Heater-to-cathode leakage in the local-oscillator tube often causes this symptom. The reduced oscillator injection voltage is the reason for the weak picture.

Picture Bent, Washed Out, and Exhibits a Hum Bar

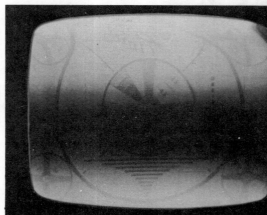

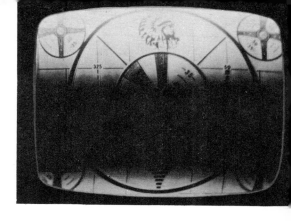

Prominent Hum Bar in Picture; Sync Lock Normal

In receivers where the sync is taken off after the video detector, this symptom is commonly caused by heater-to-cathode leakage in the video-amplifier tube.

This combination symptom indicates that hum voltage is entering before the sound take-off point. Check the video-amplifier tube for heater-to-cathode leakage. If necessary, then check preceding signal-channel tubes.

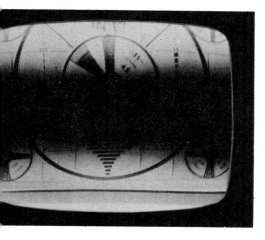

Hum in Both Picture and Sound

"Pie-Crust"
Pattern in Picture

A "pie-crust" pattern is sometimes caused by a defective horizontal-AFC, -oscillator, or -output tube. Note that some receivers have semiconductor diodes in the AFC section, instead of a tube.

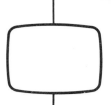

SYNC TROUBLES

Sync troubles will be considered as any defect which affects proper synchronization of the vertical and/or horizontal sweep with the transmitted signal. Thus, for our purposes, even if the defect is in the signal circuits, it will be considered a sync trouble if the sync is affected, because this is the more noticeable symptom.

In general, if both the horizontal and the vertical sync are affected, the defect is probably in the sync circuits or in the video-signal channel prior to the sync take-off point. However, if only the one—vertical or horizontal—is, the defect can still be in the video-signal channel or sync circuits, or it can be in the sweep-oscillator or AFC stages.

Another check is to try "freewheeling" the picture. That is, try critically adjusting the hold control to see if you can bring the oscillator back to the correct frequency. If so, the defect is probably in a circuit preceding the oscillator.

Many receivers develop vertical-sync instability when DC supply voltages are low. Therefore, if the symptoms are accompanied by a narrow, nonlinear, and/or dim picture, check the power-supply rectifier tube (or semiconductor rectifiers) and the damper and horizontal-output tubes.

Trouble *Page*

No Horizontal or Vertical Sync; Picture "Freewheels"
Both Horizontally and Vertically 67
No Horizontal or Vertical Sync; Strong Hum Bar in Picture 67
Picture Does Not Lock Tightly; Vertical-Hold Range Limited;
Retrace Lines Prominent 68
Vertical Sync Lock Unsatisfactory; "Hammerhead" Normal 68
Vertical Sync Lock Unsatisfactory; "Hammerhead" Weak or
Missing 69
Unstable Vertical Sync; Picture Jitters Up and Down
Continuously 69
Loss of Vertical Sync; Picture Intermittently Rolls Up a Frame .. 70

continued on following page

Trouble	Page
Picture Locks Out of Phase Vertically *(*May Drift Slowly Up or Down*)*; No Visible Hum Bar	70
Picture Locks Uncertainly in Vertical Sync; Displays "Repeats," and Often Bent; Grays Are Distorted	71
Vertical Sync Lock Unstable; Picture Negative; Contrast Poor	71
Picture Rolls Downward, Even with Vertical-Hold Control at End of Range	72
Picture Rolls Up or Down Continuously and Cannot Be Locked *(*"Freewheels"*)*; Hum Bar Prominent	72
Picture Rolls Up or Down Continuously and Cannot Be Locked *(*"Freewheels"*)*; No Hum Bar Visible	73
Unstable Horizontal Sync—Picture Jitters Horizontally	73
Horizontal Sync Unstable—Top of Picture "Pulls"	74
Horizontal Sync Unstable—Entire Picture "Pulls" and Wavers, Particularly During Commercials	74
Picture Bent and Often Jittery; Sync Lock Unstable	75
No Horizontal Sync *(*Strips Slant Upward*)*; Horizontal-Hold Control Out of Range	75
No Horizontal Sync *(*Strips Slant Downward*)*; Horizontal-Hold Control Out of Range	76
No Horizontal Sync—Horizontal-Hold Control Out of Range; Raster Narrow and Often Dim	76
No Horizontal Sync; Picture Bent and Negative	77
No Horizontal Sync; Picture Can Be "Freewheeled" with Horizontal-Hold Control	77

No Horizontal or Vertical Sync; Picture "Freewheels" Both Horizontally and Vertically

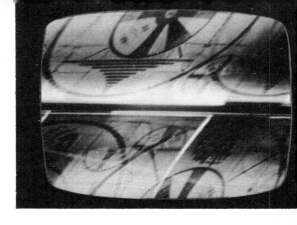

Neither the horizontal- nor the vertical-sync pulses are passing through the sync section, but the fact that picture information is displayed on the screen indicates the defective tube is in the sync section—probably the sync-separator tube.

This symptom points to a faulty tube in the video-signal channel, before the sync take-off point. Check the tubes from the RF amplifier to the sync take-off for heater-to-cathode leakage.

No Horizontal or Vertical Sync; Strong Hum Bar in Picture

Picture Does Not Lock Tightly; Vertical-Hold Range Limited; Retrace Lines Prominent

This symptom is caused by 60-cycle hum voltage, which interferes with the vertical sync and blanking pulses. In some receivers there will also be hum in the sound. If no hum bar is visible in the picture, check the local-oscillator tube first. If OK, next check each tube in turn, from the RF amplifier to the video amplifier.

Display of a normal blanking bar, or "hammerhead," between frames as shown here indicates the faulty tube is in the sync and not in the video-signal channel. In other words, the sync pulses are being fed to the sync section. (Turn up the brightness control to view the "hammerhead.") The sound will be unaffected. (Photo shows blanking-bar only.)

Vertical Sync Lock Unsatisfactory; "Hammerhead" Normal

Vertical-Sync Lock Unsatisfactory; "Hammerhead" Weak or Missing

A washed-out hammerhead means the faulty tube is in the video-signal channel, not in the sync section. That is, the sync section is not receiving normal sync pulses. The sound is usually normal, but may be weak. (Photo shows blanking-bar only.)

Unstable vertical sync is usually associated with poor sync separation and/or noise rejection. It can also be caused by faulty sync-phase inversion. Check the sync-separator, noise-canceller, and phase-inverter tubes. If they are OK, check the vertical-oscillator tube.

Unstable Vertical Sync; Picture Jitters Up and Down Continuously

Loss of Vertical Sync; Picture Intermittently Rolls Up a Frame

Loss of vertical sync is caused by defects in the same tubes listed in the previous example for unstable vertical sync. Here, however, the defect is present to a greater degree.

When the picture locks out of phase, heater-to-cathode leakage in the vertical-oscillator tube is often the cause. Lack of a visible hum bar indicates the faulty tube is not in the signal channel.

Picture Locks Out of Phase Vertically (May Drift Slowly Up or Down); No Visible Hum Bar

Picture Locks Uncertainly in Vertical Sync; Displays "Repeats," and Often Bent; Grays Are Distorted

Here the vertical-sync pulse is weak or absent in the IF amplifier because of narrow bandpass. Prior substitution of an IF tube with excessively high g_m is usually the reason. The sound is occasionally weak and noisy.

This symptom usually results from a gassy video-amplifier tube, or a gassy IF tube in receivers having high-resistance AGC systems. Sync buzz is often present in the sound.

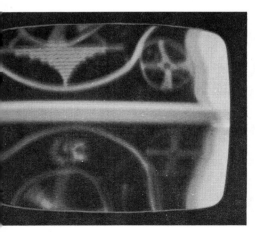

Vertical Sync Lock Unstable; Picture Negative; Contrast Poor

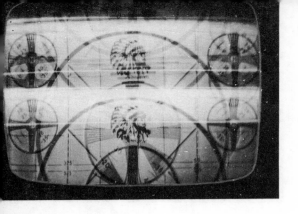

Picture Rolls Downward,
Even with Vertical-Hold
Control at End of Range

When the picture rolls downward, the vertical oscillator is running too fast. Look for a gassy vertical-oscillator tube, or one with low emission.

The hum bar in the picture indicates heater-to-cathode leakage in an RF- or IF-amplifier tube (when the sync is taken off after the video-output tube, suspect this tube also). This symptom indicates that the hum voltage is 180° out of phase with the vertical-sync pulse and has "wiped out" the pulse." In some receivers there will be hum in the sound.

Picture Rolls Up or
Down Continuously and
Cannot Be Locked
("Freewheels");
Hum Bar Prominent

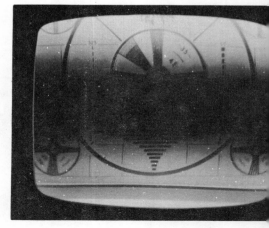

Picture Rolls Up or Down Continuously and Cannot Be Locked ("Freewheels"); No Hum Bar Visible

When the picture rolls up or down but the sound is normal and no hum bar is visible, the sync-separator tube is probably gassy. If the sync-separator tube is OK, check the sync-amplifier and noise-canceller tubes (if present). If necessary, proceed to the IF-amplifier tubes.

Picture jitter can be caused by a defective horizontal-oscillator or -AFC tube, or less commonly, by a faulty horizontal-output tube. Make a preliminary test by tapping each of these tubes and observing which one increases the jitter.

Unstable Horizontal Sync—Picture Jitters Horizontally

73

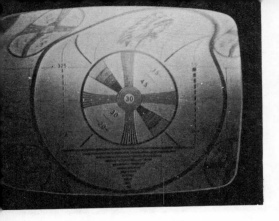

Horizontal Sync Unstable—Top of Picture "Pulls"

Pulling at the top of the picture indicates disturbance of the horizontal-sync function following passage of the vertical-sync pulse. Check the sync-separator and sync-amplifier tubes (if present). Also look for a weak AGC keyer or amplifier tube (if present), and gassy IF tubes.

This symptom indicates marginal sync separation, or overloading in a single circuit prior to sync take-off. Check the sync, AGC, and IF tubes—and also the video-amplifier tube if prior to sync take-off. When the faulty tube is in the signal channel, sync buzz will often be heard, too.

Horizontal Sync Unstable—Entire Picture "Pulls" and Wavers, Particularly During Commercials

Picture Bent and
Often Jittery; Sync
Lock Unstable

Heater-to-cathode leakage in the horizontal-AFC, sync-separator, or sync-amplifier tube (if present) commonly causes this picture symptom. If the heater-to-cathode leakage is in the sync-separator tube, a dim hum bar can sometimes be seen in the picture (reflected symptom).

The strips slope uphill to the right because the horizontal-oscillator is running too slow. Check the horizontal-AFC tube (or semiconductor diodes). If OK, proceed to the sync-phase inverter tube, and, in some receivers, to the noise-canceller tube.

No Horizontal Sync
(Strips Slant Upward);
Horizontal-Hold Control
Out of Range

No Horizontal Sync
(Strips Slant Downward);
Horizontal-Hold Control
Out of Range

The strips slope downward to the right because the horizontal oscillator is running too fast. Gas or low emission in the horizontal-oscillator tube can cause this symptom, as can a faulty horizontal-AFC tube (or semiconductor diode). In this and the previous illustration, the pattern is locked into vertical sync, showing that the sync-separator tube is not a suspect.

A narrow and dim raster points to low emission in the horizontal-oscillator tube. The low emission reduces the oscillator grid-current flow, resulting in loss of horizontal sync.

No Horizontal Sync—
Horizontal-Hold Control
Out of Range; Raster
Narrow and Often Dim

No Horizontal Sync;
Picture Bent and
Negative

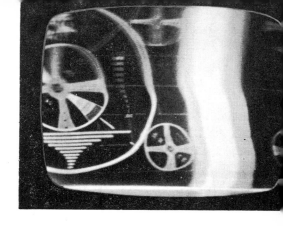

No emission in the keyed-AGC tube may cause this symptom. The lack of bias output overloads the IF (and sometimes the RF). If the keyed-AGC tube is OK, look for gassy IF tubes. Some receivers develop this symptom when the video-detector semiconductor diode is defective.

This symptom shows that the horizontal-sync pulses are being stopped in the horizontal-sync section. Check the horizontal-AFC tubes for semiconductor diodes). If the picture locks out of phase, look for heater-to-cathode leakage in the horizontal-AFC tube.

No Horizontal Sync;
Picture Can Be
"Freewheeled" with
Horizontal-hold Control

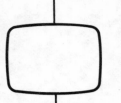

SIZE AND LINEARITY TROUBLES

Size troubles include all defects which affect the height and width of the raster. Linearity troubles are those which produce compression or expansion of a portion of the raster; they are usually caused by defects in the sweep circuits. However, watch out for interacting troubles, and keep the functional layout of the receiver in mind. Since the boost voltage is usually applied to the sweep output tube (and often to the oscillators), do not overlook the possibility of incorrect boost voltage due to a defective damper tube. The low-voltage rectifier also may cause insufficient height and width.

Keep in mind that, although these symptoms are usually produced by defective tubes, the defect may be in the associated circuitry and further troubleshooting will be required.

Trouble	*Page*
No Vertical Sweep (Picture Collapsed to a Horizontal Line)	80
Excessive Height; Raster Dim and Decentered	80
Insufficient Height; Picture Compressed at Bottom	81
Bright Horizontal Line (Foldover) at Bottom of Raster	81
Vertical Nonlinearity—Picture Expanded at Top and Compressed at Bottom	82
Vertical Nonlinearity—Picture Flattened at Top and Bottom	82
Insufficient Height; Picture Rolled Up	83
Picture Compressed on Right Side	83
Picture Compressed and Folded Over on Right Side; Raster Dim	84
Picture Distorted and Compressed on Left Side	84
Picture Compressed on Left Side; No Shading or Other Distortion	85
Bright Vertical Strip (Foldover) at Right Edge; Raster May Be Dim	85
Bright Vertical Strip (Foldover) at Left Edge; Raster May Be Dim	86

Trouble	Page
Excessive Height and Width; Raster Dim; Picture Often Appears Filled Up	86
Insufficient Height and Width; Raster Dim; Picture Often Appears Filled Up	87
Insufficient Height and Width; Poor Linearity; Raster Dim	87

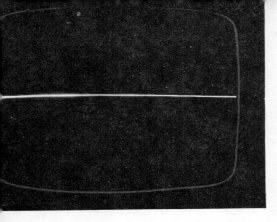

No Vertical Sweep
(Picture Collapsed to
a Horizontal Line)

Either the vertical-oscillator or vertical-output tube is dead. (In some receivers, both functions are combined in the same tube.)

Here is an example of a symptom reflected from the horizontal to the vertical section. Low emission in the horizontal-output tube causes a decrease in the high voltage, in turn causing the increased height (the width is not increased because of the lower horizontal-sweep voltage). The dim raster is your clue to the reflected trouble.

Excessive Height; Raster Dim and Decentered

Insufficient Height;
Picture Compressed
at Bottom

This symptom is caused by low emission in the vertical-output tube. A gassy vertical-output tube will cause a similar picture symptom.

In some receivers this symptom is caused by a weak vertical-output tube, the foldover line appearing when the vertical-height control is advanced. In other receivers, a weak damper tube is often the reason.

Bright Horizontal
Line (Foldover) at
Bottom of Raster

**Vertical Nonlinearity—
Picture Expanded at
Top and Compressed
at Bottom**

This form of vertical nonlinearity is caused by heater-to-cathode leakage in the vertical-output tube. The leakage drains out DC cathode bias; it also injects 60-cycle hum voltage into the sawtooth drive, and increases the nonlinearity.

The nonlinearity is the result of low emission in the vertical-output tube, and the flattening at the top and bottom comes from attempted correction by readjustment of the vertical-linearity and height controls.

**Vertical Nonlinearity—
Picture Flattened at
Top and Bottom**

Insufficient Height; Picture Rolled Up

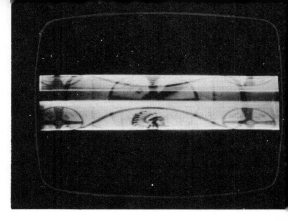

This symptom occurs when there is heater-to-cathode leakage in a combination vertical-oscillator–output tube. The leakage stops the sawtooth oscillation, but injection of 60-cycle hum voltage at the cathode causes limited vertical deflection and a "rolled-up" display.

Picture compression at the right is caused by gas in the horizontal-output tube. The gas lowers the grid impedance and distorts the top of the drive waveform.

Picture Compressed On Right Side

Picture Compressed and Folded Over on Right Side; Raster Dim

This symptom, also caused by a gassy horizontal-output tube, indicates a higher gas content than in the previous example.

Low emission in the damper tube will cause this symptom in many receivers, because the damper tube controls scanning on the left-hand portion of the picture.

Picture Distorted and Compressed on Left Side

Picture Compressed on
Left Side; No Shading
or Other Distortion

Another symptom of faulty damper operation—it is commonly caused by leakage or traces of gas in the damper tube. If the damper tube is in good condition, then look for faulty circuitry in the damper section.

Foldover at the right side of the raster is caused by gas in the horizontal-output tube. In some receivers this symptom is accompanied by horizontal nonlinearity.

Bright Vertical Strip
(Foldover) at Right Edge;
Raster May Be Dim

Bright Vertical Strip (Foldover) at Left Edge; Raster May Be Dim

When the raster is folded over at the left, the damper tube is defective. If the tube is in good condition, the fault is in the damper circuitry.

When a set exhibits these symptoms and the linearity remains normal, look for a high-voltage rectifier tube with low emission. If it is in good condition, suspect the high-voltage circuitry.

Excessive Height and Width; Raster Dim; Picture Often Appears Filled Up

Insufficient Height and
Width; Raster Dim;
Picture Often Appears
Filled Up

Low emission in the low-voltage rectifier tube, or faulty semiconductor rectifiers, are to blame here.

Some receivers also display prominent nonlinearity (vertical and/or horizontal) when the plate and screen supply voltages are low (due to low emission in the low-voltage rectifier tube, or to faulty semiconductor rectifiers). Otherwise, this symptom is the same as the previous one.

Insufficient Height and
Width; Poor Linearity;
Raster Dim

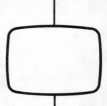

RASTER TROUBLES

The troubles covered in this section include those which affect the raster itself, whether or not there is a picture. Such troubles include spots, lines, and other distortions on the screen, even on vacant channels. Size troubles could also be considered raster troubles, but they were covered in the previous section because of their interrelation with linearity troubles, which are apparent only when a picture is being shown.

The symptoms in this section are usually caused by certain tubes in the receiver, irrespective of its functional layout.

Trouble	Page
Pincushioned Raster	89
Small Dark Spot in Center of Raster	89
Large Dark Spot in Center of Screen	90
Dark "X" in Center of Raster	90
Vertical Line(s) at Left Edge of Picture	91
Ragged Vertical Line at Extreme Left Edge of Raster	91
Dark Splotches (Snivets) at Right Side of Raster	92
Streaked Raster	92
No Raster; Sound Normal	93
No Raster; No Sound	93

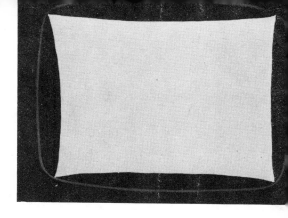

Pincushioned Raster

Pincushioning (curvature at the edges) occurs when the yoke and the picture tube are not properly matched. However, if the receiver uses anti-pincushion magnets and they are misadjusted, pincushioning may result from the magnetic fields in the vicinity of the picture tube. To correct, use a matching tube (or yoke), or adjust the anti-pincushion magnets.

A small dark spot in the center of the raster is a screen burn, caused by an intense bright spot remaining on the screen for several seconds each time the receiver is turned off. The only cure is a new picture tube. (As a preventive measure if a spot remains on the screen, make it a practice to turn down the brightness control before turning off the set.)

**Small Dark Spot
In Center of Raster**

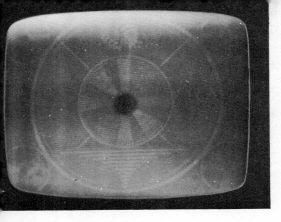

**Large Dark Spot
In Center of Screen**

An ion burn has ruined this picture tube, and all you can do is replace it. Make certain the ion trap on the new tube is correctly adjusted, to avoid the same trouble later.

Another example of an ion burn. The "X" burn is more common in rectangular tubes, and the round burn in round tubes. Again, make sure the ion trap is correctly adjusted after replacing the tube.

**Dark "X" in
Center of Raster**

Vertical Line(s) At Left Edge of Picture

Called Barkhausen lines, they can often be minimized or eliminated by replacing the horizontal-output tube. In some receivers you may have to try several tubes before the lines are eliminated.

These are "spook" lines and they can sometimes be minimized or eliminated by replacing the damper tube. In some receivers, several tubes may have to be tried before one is found which will correct the trouble.

Ragged Vertical Line At Extreme Left Edge of Raster

Dark Splotches (Snivets) At Right Side of Raster

Snivets can also be minimized or eliminated by replacing the damper tube. As before, several tubes may have to be tried.

Streaks or flashes in the raster are usually caused by arcing in the damper, horizontal-output, or high-voltage rectifier tubes. If these tubes are OK, check the vertical-output and the picture tube.

Streaked Raster

NO RASTER; SOUND NORMAL

A completely dark screen and normal sound indicates that the receiver circuits are working up to the sound take-off point and that the sound section also is operating. Check the horizontal-oscillator, horizontal-output, damper, and high-voltage rectifier tubes. If they and their circuits are operating properly, the picture tube is probably defective. Note, however, that many other circuit defects can result in loss of high voltage—which in turn causes loss of raster.

NO RASTER; NO SOUND

If the set is completely dead—that is, none of the tubes are lit, look for a blown power fuse in the receiver or, in the case of series-string heaters, a tube with an open heater. If all tubes are lit, look for a defective rectifier tube (or semiconductor rectifier) in the power supply.

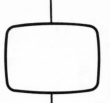

SOUND TROUBLES

In this section, troubles which affect primarily the sound reproduction are discussed. No picture-tube displays are included; troubles which affect both the picture and the sound were displayed in the previous sections. First examine the picture, to see if it is affected and how. If so, look in the applicable section and find the possible causes. Otherwise, look under the symptom in this section.

Unless otherwise specified, the picture is normal in receivers exhibiting the following sound symptoms.

Sound defects include hum, a soft 60-cycle tone; buzz, a harsh and rasping 60-cycle tone; and garbled, mushy, or boomy reproduction. Hiss is an interference voltage which has entered the sound channel.

Trouble	*Page*
Sound Weak or Absent	95
Intermittent Sound	95
Hum in Sound	95
Sync Buzz in Sound	95
Hiss in Sound	95
Ringing or Howling in Sound; Streaks May Be Visible in Picture	96
Distorted Sound	96

SOUND WEAK OR ABSENT

Check the audio-output, audio-amplifier (if present), audio-detector, and sound-IF amplifier tubes—they may have low emission or are dead. If there is a raster but no picture, check the video-amplifier, video-detector, IF- and RF-amplifier, and oscillator-mixer tubes. In some receivers, a dead audio-output tube will cause loss of picture. If there is no raster, check the low-voltage rectifier tube (or semiconductor rectifiers).

INTERMITTENT SOUND

Loose elements or intermittent leakage in the sound-IF amplifier, audio-detector, audio-amplifier, or audio-output tube can cause the sound to come on and off at intervals.

HUM IN SOUND

When the picture is normal and 60-cycle hum is heard from the speaker, check all tubes beyond the sound take-off point for heater-to-cathode leakage or shorts. These include the sound-IF amplifier, audio-detector, audio-amplifier, and audio-output tubes.

SYNC BUZZ IN SOUND

When sync buzz is not accompanied by an overloaded picture or loss of sync, check the sound-IF amplifier, sound-limiter (if present), and audio-detector tubes.

HISS IN SOUND

A gassy audio-amplifier or audio-output tube can generate a hiss in the sound. A defective outer coating on the picture tube will produce small arcs or corona, another cause of hiss.

RINGING OR HOWLING IN SOUND; STREAKS MAY BE VISIBLE IN PICTURE

This symptom is caused by a microphonic tube in the signal circuits. To find it, tap each tube in the signal channel, from the RF amplifier to the audio amplifier, one by one. A microphonic tube will emit a loud ringing noise in the speaker when tapped.

DISTORTED SOUND

When the sound is distorted but the picture is normal, the audio-amplifier or -output tube probably has low emission or is gassy.